SIMPLY PAINT

50 CREATIVE IDEAS FOR IMPROVING YOUR HOME

LINDA BARKER
PHOTOGRAPHY BY LIZZIE ORME

SIMPLY

PAINT

50 CREATIVE IDEAS FOR IMPROVING YOUR HOME

LINDA BARKER
PHOTOGRAPHY BY LIZZIE ORME

ANAYA PUBLISHERS LTD
LONDON

For Chris and Jessica

First published in Great Britain in 1993 by

Anaya Publishers Ltd
Strode House, 44-50 Osnaburgh Street, London NW1 3ND

Copyright © Anaya Publishers Ltd 1993

Series Editor   Janet Donin
Designer   Jerry Goldie
Photographer   Lizzie Orme
Assisted by   Sussie Nielsen
Stylist   Linda Barker
Assisted by   Simon Matthews
Sally Barker
Jacket design by   Watermark

British Library Cataloguing in Publication Data
Linda Barker
Simply Paint. – (Simply Series)
I. Title   II. Series
747

ISBN 1–85470–084–7
Typeset in Great Britain by Bookworm Typesetting, Manchester
Colour reproduction by Scantrans Pte, Singapore
Printed and bound in Italy by Officine Grafiche de Agostini

Contents

Introduction
7

INTRODUCTION

I'm sure you'll have great fun adding colour and paint techniques to your home and favourite possessions.

I love decorating my home, don't you? Whenever I'm feeling bored I pick up a brush and paint something. The ceiling of my dining room has stars painted on a bruised lilac background and in the living room there's a stormy sky ceiling! They were my attempt at getting rid of the cracks! My favourite piece of furniture is an old desk, found in a secondhand shop, which I painted to look like a grand architectural building. The main elements in all these projects are colour and paint, because for me painting is very refreshing. I hope you enjoy it too.

I think most of us do have this basic and irresistible urge to decorate our house. To personalise it, to stamp our taste on it and make it home. From a new lick of paint on the woodwork, to grandiose schemes that would put the Sistine Chapel to shame, the way we use pattern and pigment in our homes vividly reflects our personality. An inspired choice of colour, bold strokes with the brush and some deft dabbing with a bathroom sponge can transform any drab room into a riot of fabulous paint effects. It needn't cost very much and it certainly doesn't require a fistful of technical qualifications, just lots of enthusiasm and a little patience. Nowadays, there's really no need to spend your life staring at magnolia walls and white glossed woodwork when there are so many paints in glorious colours. And purchasing a few simple tools will allow you to transform just about anything, from a little box to a spare bedroom, into a unique example of your skills and imagination. There are, of course, numerous ways in which you can decorate your home, but paint, above all, offers quick, simple and often dramatic results without having to remortgage your house! In SIMPLY PAINT I have selected 50 varied techniques that will allow you to use paint to its full potential. Most are easier to perfect than you would have guessed, and those that do require a little more perseverance are well worth the effort. Once you have completed this book and mastered all the skills it contains, that's just the beginning! From then on you can develop your own variations and adaptations, conjure up original tints, mix your own glazes, transform objects I've never thought about, and even go on to develop tricks of your own. You're far more likely to run out of unpainted walls and furniture before you exhaust all the colours, materials and methods at your disposal!

Paint offers limitless possibilities and in this book I hope to release any inhibitions you may have about using it. Fancy crackling the cupboard? No problem! Sponging the spare room? Read on! Verdigris walls and a tartan chest of drawers? Pick up your brushes now! I've always believed that colour and finish in decoration should be fun and exciting, and even if you do make a mistake by choosing a stronger shade than you had envisaged, I'll show you how to tone it down without having to start all over again.

In SIMPLY PAINT I'll introduce you to the fantastic versatility of the medium. The easy to follow step-by-step instructions will enable you to master techniques you have previously only admired in design books and magazines. By the way, there's a glossary of unfamiliar terms for readers worldwide on page 112. I'll also show you how to transform pieces of junk and ordinary furniture into beautiful pieces that will make you justifiably proud. And if you feel like tackling whole rooms, but don't have the time or money for a complete rethink, you'll find there are paint treatments that can make use of existing wall colours too.

Creating your own, individual look for a room, or painting a piece of furniture is immensely rewarding. Hopefully, by following my tried and tested short cuts, you will soon be achieving fabulous results that will be the envy of all your friends. Above all have fun with these ideas and don't be afraid to experiment. Practise the techniques on a piece of cardboard or plywood before you commit yourself to the real thing. Rehearse the more complicated skills until you are confident and don't be afraid to play around with a variety of colour combinations – pigments of your imagination! Most importantly, enjoy the painting, then just sit back and wait for the compliments. After all, once you've removed stubborn paint from under your fingernails, hammered your crusty brushes into their soft, original condition and finally cleared your head of the turpentine fumes, you'll thoroughly deserve them!

Preparation and Materials

Before painting any surface it is important to consider the preparation. This, I'm afraid, is usually quite time consuming, involving several stages to achieve a perfect finish before any surface decoration can take place. But it's a procedure well worth your time and effort. After all, what artist in their right mind would consider starting their next masterpiece on a scabby old canvas, pockmarked with blemishes and congealed debris?

Cleaning A general household detergent will remove most traces of grease which would otherwise resist the paint.

Filling Deep scratches, cracks and holes should be filled using an appropriate filler. For small projects choose a ready mixed filler that's quick and easy to use. For larger projects it may be more cost effective to mix your own. Always overfill the crack or hole as the filler will shrink slightly, then sand to a smooth surface when dry. If the crack or hole is quite deep, apply the filler in two or three coats, allowing it to dry between each application.

Sanding There are many grades and types of sandpaper so it's important to use the correct one for the job, depending on whether you are sanding down fillers or preparing the surface to accept a primer. The information on the packaging will usually advise you on the right choice. A sanding block, around which to wrap your sandpaper, is invaluable for use on smaller projects but you'll probably find an electrical sander more useful for larger areas and wide horizontal surfaces. Always try to sand the surface using a circular motion.

I painted this chest of drawers to look like a Venetian villa. It may look complicated but as with all the projects in this book it is very simple. Just follow my step-by-step instructions for professional results.

Priming Primers are important to protect and seal the surface from moisture and dirt. Again the surface to be painted will govern the type of primer you should use. Primer is usually white and available in three types: water, oil and alcohol based. Decide on the type of paint finish you wish to use, then choose the corresponding primer. Oil and water based paints will adhere to an alcohol or water based primer but only use an oil based paint over an oil primer.

Base coat The base coat will provide a smooth layer upon which to paint and should, in many cases, be applied in two coats. It is important to work in a dust-free room at this stage to ensure that your paint finish has a smooth, blemish-free base. For a professional finish, thin the first coat with a small amount of the appropriate solvent (approximately 10/15% water for water based paints or white spirit for oil based paints). Allow the first coat to dry completely before applying the second coat. This should also be thinned slightly with the appropriate solvent (approximately 5/10%).

Remember that no two jobs are the same and these steps are given only as a general guideline. Some surfaces will simply require a quick rub with fine sandpaper before base coating, others will require considerably more treatment. Unfortunately there are no short cuts to success! A badly prepared surface will always show through subsequent layers, so time spent on the preparation stage will produce a more satisfying result. It's a daunting prospect I know but at least the furniture is going to last.

Depending on the item to be painted it may be necessary to use chemical strippers to remove old layers of paint. Always use these following the manufacturers' instructions and do work in a well-ventilated room as this can be smelly work.

Equipment

The equipment you will need for your paint projects shouldn't cost a fortune, although it's always advisable to buy the best quality products you can afford. The following list should equip you with the necessary tools to start your basic kit.

Brushes A good selection of brushes in various sizes is essential, but be discerning; don't buy the cheapest. There's nothing quite so infuriating as a balding brush leaving stray bristles on your newly painted surface! Some brushes are designed to use with a particular paint and although these are useful I don't find them strictly necessary. However, it's common sense to use a wide brush to cover larger areas and a narrow brush for more detailed work. A thicker brush will hold more paint than a thinner one. The most important advice I can give you is to clean your brushes thoroughly between each job.

For certain paint techniques it is easier to use a specialist brush specifically designed for that treatment, but these can be expensive. You can often get a good enough result by adapting a general household brush. For example the dragging technique can almost always be achieved using a long haired paint brush rather than having to buy an expensive dragging brush. However for some decorative techniques there is sometimes no substitute. Stippling can really only be successfully achieved with a stippling brush. So if you plan to use a particular technique quite often, it may be worth investing in the relevant brush.

Dust sheets These are always useful to have as part of your basic equipment, to protect furniture or floors from splatters or spills. I buy cheap cotton sheets on special offer, which do the job just as well for a fraction of the cost.

Cotton cloths These can often be purchased on a long roll and are brilliant for cleaning brushes, surfaces and all those nasty spills. Choose lint-free cloths as these won't leave fluff over everything. Some paint finishes use cloths rather than brushes, so it's useful to have a good supply. Cheesecloth and hessian cloth are used for cloth-distressing techniques.

Sponges Cellulose decorator sponges are helpful when preparing surfaces and can also be adapted for particular paint finishes if you can't obtain a large natural sea sponge.

Containers Paint kettles or plastic containers are essential when thinning paint or mixing glazes. Paint trays in various sizes ease some applications and jars with screw-top lids are excellent for storing paint.

Tapes Masking tape and low tack masking tape make it easier to paint straight lines and are useful to protect your work.
Other pieces of equipment like pencils, tape measures and craft knives are all useful additions to your basic kit, which no doubt will be growing all the time. And remember your kit will last longer if you clean brushes and tools diligently after each job and dry them thoroughly before putting them away.

Painting

Once you have all the materials together and the wall or piece of furniture has been prepared and the base coat applied, you can at last begin to paint and the fun should really start!

Throughout this book I've put together projects that use both oil and water based paints. Sometimes I've used paint directly from its can on to the surface, but more often than not I prefer a softer, more subdued quality. For these finishes I have either thinned the paint with the relevant solvent (water for emulsions, white spirit for oil based paint) or I have used transparent glazes. These are available with both oil and water bases and can be tinted with an appropriate paint. Depending on the type of finish you want, any paint can be thinned using the glazes. Emulsion and eggshell (satinwood) paint can have glazes added to them to give a semi-transparent quality, but tinting the glaze with pure artist's colours will produce a transparent finish. Where it is necessary I have given the proportions of paint to glaze and also which type of paint to use. You will find the name of the paints for each project at the back of the book.

Finally I hope you enjoy reading and working through this book and have fun because that's really what SIMPLY PAINT is all about.

CHAPTER 1

PAINT TECHNIQUES

You've all heard about paint finishes and no doubt seen wonderful examples of some of the better known techniques, but I bet you don't realise just how simple it can be to actually paint this way. For instance, would you believe it could be so easy to achieve the wall finish on the front cover? Yet all you need is two shades of terracotta paint, a splash of water and the biggest flat brush you can find! Follow the same technique on a smaller scale and you can create the glorious colour-washed bathroom. On the following pages I'll also show you how to use a rubber comb to pattern a picture frame, how to pattern your own china and even how give your doors a smart new dragged look. There is really no end to the wonderful effects you can create with a little know-how and this chapter should give you all the basic knowledge to have a go!

Sponged
Walls

This is one of the simplest paint techniques as large areas can be worked quite quickly, so it's an ideal finish for the beginner. I have used three colours sponged over a base coat to get this attractive, soft, cloudy appearance.

MATERIALS

natural sea sponge

emulsion paint – base colour

3 shades emulsion paints – sponging

paint kettles

shallow paint tray

10cm/4in wide paint brush

Hints

For best results use a large flat sea sponge. If only round ones are available, slice in half and use the cut surface to apply the paint.

1 Prepare the wall by washing down and filling any surface cracks. Paint with two coats of your base colour. Mix each of the three shades to be sponged with one part paint to one part water in a paint kettle.

2 Put a small amount of the darkest colour into the tray and dip the sponge lightly into the paint. Don't overload with paint. Dab lightly onto the wall, varying the imprint by turning the sponge slightly after each mark has been made.

3 Wash and dry the sponge before each new application. Sponge on the second colour in the same manner, although this time you can apply the paint a little more patchily, filling the gaps and overlapping the first colour.

4 Finally repeat the sponging process with the lightest colour. I think it's a good idea to keep standing back from the wall to check the overall look before adding more of this last colour to achieve the desired effect.

Sponged Coffee Table

If you can't find a sea sponge, adapt a cellulose decorator's sponge by cutting it into an oval shape and randomly cutting out varying sized holes to resemble those found in a natural sponge. Use oil based paint for a tougher finish as this type of table will probably be in constant use.

Materials

sea sponge or cellulose sponge

white eggshell – base coat

3 colours eggshell paints – sponging

shallow paint tray

white spirit

paint kettle

paint brush

sandpaper

newspaper

Hints

If you feel that you have lost the cloudy effect by the time the last coat has been applied, try sponging over with a little of the darkest colour. Use sparingly until you have the desired finish.

1 Prepare the table by removing any old paintwork and sanding to a smooth finish. Apply the base coat to the furniture. Mix each of the paints for sponging in a paint kettle, using one part paint to two parts white spirit.

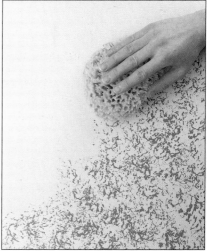

2 Using the darkest paint colour first press the sponge lightly into the paint. Use a dabbing motion to apply the colour to the surface of the table. You may need to work off a little of the colour onto newspaper first to achieve a suitable imprint.

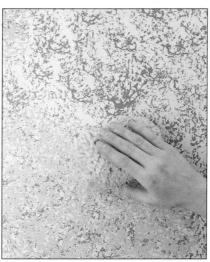

3 Use warm soapy water mixed with white spirit to clean the sponge between each paint application. Repeat the sponging process with the second colour building up a patchy surface. Be careful not to sponge too heavily at the edges of the table and on the legs.

4 With a clean dry sponge finally apply the last colour. As this is the palest shade it is used to give a soft, cloudy appearance, so apply lightly and sparingly at first. Check the overall effect regularly before adding more colour as needed.

Sponged Ceramics

Use commercially prepared ceramic colours which are "thermo-hardening", meaning they should be baked in the oven. This gives the colours a very hard and scratch-resistant surface, although the ceramics should not be washed in the dishwasher.

MATERIALS

small sea sponge

ceramic colours

scissors

pencil

brown paper

small plate

double-sided tape or spray adhesive

saucer

pre-cut stencil

masking tape

1 Clean the surface of the ceramics with soapy water to remove any dirt or grease. Before decorating a plate, prepare a mask from brown paper. Place a smaller plate, the same diameter as the centre of your plate, on the paper, draw around, then cut out.

2 Secure the mask to the plate with double-sided tape or spray adhesive. Pour a little ceramic paint into a saucer and lightly press the sponge into the paint. Dab the sponge over the unmasked area of the plate, turning it each time to create an irregular pattern.

3 Apply subsequent colours with a clean, dry sponge. When the border is dry, position the first stencil layer on the plate. Secure with masking tape. Again pour the first ceramic colour to be used into a saucer and follow the same sponging technique.

4 Allow the ceramic paint to dry thoroughly before positioning any subsequent layers of the stencil and repeating the stencilling. If you want a denser colour on some areas of the pattern, use the sponge to build up further layers of colour.

5 Again let the colour dry before positioning the last stencil layer and applying the remaining colour as before. When this paint is dry, put the plate in the oven. Follow the paint manufacturers' directions and bake for the required period of time.

Colour-rubbed Storage Box

This is a simple way to liven up any plain storage box. For an interesting effect I have colour rubbed the scallop shells so only a little paint is held in the recesses, allowing the attractive natural markings of each shell to be seen.

MATERIALS

white oil based primer

coloured emulsion paint

scallop shells

fine sandpaper

paint brush

paint kettle

cotton cloth

newspaper

all-purpose glue

1 Paint the frame of the box using white primer. I used two layers of primer here to ensure a good, even coverage. Rubbing down with fine sandpaper between the layers gives a smooth finish.

2 Mix your chosen emulsion colour with an equal quantity of water in a paint kettle and apply liberally to the surface of each shell. Emulsion paint left from painting your bathroom could be used here.

3 Use a dry cotton cloth to remove most of the wet paint from the shells. The small amount of paint that is retained is sufficient to colour the shell while allowing the natural texture to show through. Glue dry shells to the bathroom storage box.

Colour-rubbed Linen Basket

The scallop shell detail glued to this ordinary linen basket transformed it into something really special and allowed me to use two types of colour rubbing, one on the basket weave and another on the shells.

MATERIALS

scallop shells

white oil based primer

coloured emulsion paint

medium grade wire wool

fine sandpaper

long haired brush

paint brush

paint kettle

cotton cloth

newspaper

all-purpose adhesive

Hints

You could use oil instead of water based paints if your bathroom is particularly humid or if the basket is placed near to the bath. But remember this will have to be diluted with white spirit.

1 Base-coat the linen basket with white acrylic primer. Use a narrow but long-haired brush to help push the paint right into the basket weave. If you haven't got the patience to do this, you can cut corners and use a spray-on acrylic primer.

2 When the base coat is thoroughly dry use the same clean brush to apply the coloured emulsion and leave to dry. Don't worry if the paint dribbles through to the inside. You can line the basket with fabric afterwards for a really professional finish.

3 Using wire wool carefully rub away patches of the emulsion paint. You may want to wear rubber gloves as wire wool fibres are prickly. Use enough pressure to remove the paint without stripping back to the wicker and follow the weave of the basket.

4 Use a paint brush to dust away the particles of wire wool. If the basket needs a little highlighting, dip the tip of a dry brush into some white emulsion and use a light, flicking movement of your wrist to transfer the paint onto the raised areas of the basket.

5 The edges of scallop shells are often quite sharp so rub them with a little sandpaper before painting. Dilute the emulsion with 50% water in a paint kettle. Place the shells on newspaper, or hold carefully and apply the paint to each shell. Allow to dry.

6 Use a dry cotton cloth and rub the shells quite firmly to remove most of the colour. I think it's rather attractive to see some of the natural markings on the shells. Use strong adhesive to stick the shells firmly in position on the lid of the basket.

Combed Picture Frame

This is a wonderfully simple way to bring a new lease of life to any battered old picture frame you may have decided to throw out. If you find it difficult to obtain a commercially made comb, make your own by cutting one from cardboard, but varnish it before using to stop it from becoming soggy.

MATERIALS

white oil based primer

coloured emulsion paint – base coat

coloured emulsion paint – top coat

emulsion glaze

paint brush

paint kettle

decorator's comb

Hints

If you don't have any emulsion based glaze, it is possible to mix the paint with an equal amount of PVA glue, which is more readily available and gives a similar effect.

1 Prepare the frame with a white acrylic primer and two coats of your chosen base colour. The combing technique reveals quite a lot of the base colour, so consider your two colours carefully. Allow to dry thoroughly.

2 In a paint kettle mix equal quantities of the coloured emulsion top coat with the emulsion glaze. Apply an even coat all over the base colour. You will find that this glaze mix remains wet for some time.

3 Use your comb to pull through the glaze, creating a pattern. Vary the pattern by zig-zagging the comb or twisting it to make a number of different effects. I prefer to experiment with designs on a piece of cardboard before committing myself to the project.

Splattered Picture Frame

This method of splatter painting is sometimes known as Cissing. For this picture frame I have splattered over a silver base coat as this gives a wonderfully luminous quality to the finished frame.

MATERIALS

shellac varnish

artist's silver powder

2 tubes of artist's oil colours

white spirit

paint brush

polyurethane varnish

paint kettle

sandpaper

ruler

Hints

If you don't intend to use this paint finish often, shellac varnish may be an expensive buy, so use a solvent-based, silver paint for the base coat instead.

1 Prepare the frame by sanding down the wood to a smooth finish if necessary. Mix a little of the silver powder with the shellac varnish to form a thin base coat. When you apply this to the frame it will produce an interesting opaque finish. Allow to dry.

2 Paint one coat of polyurethane varnish over the silver top coat. This is an unusual, though essential, application as it forms the basis for the overall splattered effect. Do not let it dry out completely before progressing to the next step.

3 Thin the 2 oil colours individually with a little white spirit and dab each colour on to the frame so they sit on top of the varnish. Don't worry if the colours touch and start to run together – that's exactly what they should do!

4 Dip the tip of the paint brush into white spirit and splatter the solvent on to the frame by striking the brush against a ruler. The painted surface will now open up, forming unusual pools of colour. Leave to dry horizontally.

Colour-washed Tongue and Groove Panelling

Colour washing gives a stunning Mediterranean feel to any room or surface so don't be afraid to use the brightest colours you can find.

MATERIALS

2 oil based paint colours

white spirit

paint kettle

5cm/2in paint brush

knotting sealer

Hints

As an alternative treatment it is possible to create this effect using thinned emulsion paints washed over a prepared emulsion base coat. For the best results choose a base colour that's similar to the glaze. You will need to varnish it when finished.

1 On new panelling it is important to let the natural wood grain show through, so there is no need to prepare with a base coat, but it is essential to seal any knots in the wood with the relevant preparation to prevent any wood resins from seeping through the paint.

2 Dilute the two colours to be used with an equal quantity of white spirit in a paint kettle. Apply the darkest colour directly on to the panelling first. Brush the paint outwards so the colour is stronger in some areas and lighter in others.

3 When the first coat is completely dry apply the second colour in the same way. Leave stronger patches of colour in some areas but work the paint well into other areas for a softer look. Take care to avoid making an obvious pattern.

Ragged Chair

You'll enjoy the almost instantaneous finish you get with this technique which I find just as easy to master as sponging.
With very little effort you can achieve really professional results.

MATERIALS

coloured oil based paint – base coat

white oil based paint – top coat

transparent oil glaze

paint brush

cotton rag

white spirit

sandpaper

paint kettle

Hints

Other materials can be used to remove the glaze such as plastic or paper bags, lightweight muslin or heavyweight hessian. Each material will give slightly different effect.

1 Prepare the chair by removing any old paint, then sanding the surface to a smooth finish. Paint with two coats of your chosen base colour. Allow to dry thoroughly and sand down with a fine grade sandpaper between each coat.

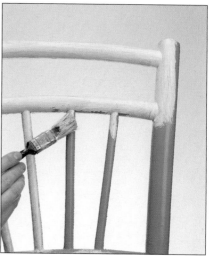

2 Mix equal quantities of the oil glaze with the white oil based top coat in a paint kettle. Paint the surface of the chair with a thin layer of this glaze mix. Be careful not to allow the paint to build up in the corners.

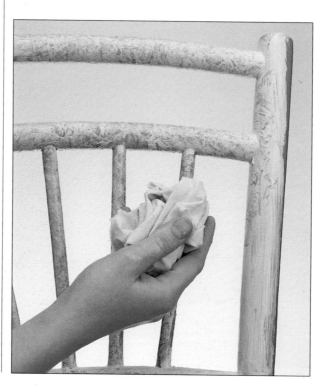

3 Loosely crumple the rag and press into the wet glaze. As you lift off the cloth you remove some of the glaze. Repeat with swift movements all over the chair. When the cloth is covered with glaze, shake it out and recrumple before continuing to complete the ragging.

Rubbed Chair

This is a beautiful way to give a softly aged look to what is actually a strong base colour. It's a technique that's particularly good for furniture to be used in a country setting.

MATERIALS

coloured oil based paint – base coat

white oil based paint – top coat

transparent oil glaze

paint brush

cotton cloth

white spirit

paint kettle

sandpaper

Hints

This technique works well on most carved wood areas and is particularly effective when used on an ornate dado or picture rails.

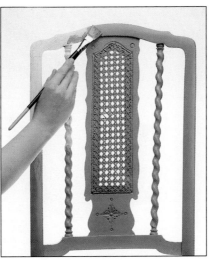

1 Prepare the chair by removing any old paint and sanding down to a smooth finish. Prime if necessary. Paint with bold base colour. Furniture with an ornate carving to hold the glaze can be even more dramatic.

2 Mix together equal quantities of the glaze and white top coat in a paint kettle. When the base colour is thoroughly dry, apply the mixed glaze colour evenly to your chosen piece of furniture.

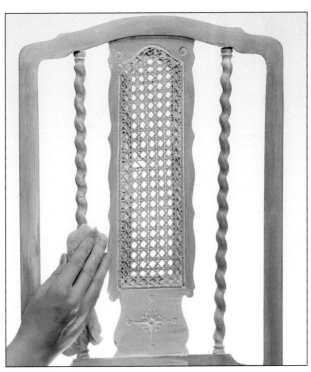

3 Allow the glaze to dry a little before you start to remove it. Use a cotton cloth folded into a smooth pad and lightly stroke off the glaze to reveal the base colour. The glaze will be held in any recesses to give a distinctive finish.

Mutton-clothed Salt Cellar

This technique can be used on many flat surfaces – even on walls, particularly where a country look is required – as it forms a cloudy appearance over a large area.

MATERIALS

oil based primer

white oil based paint – base coat

coloured oil based paint – top coat

transparent oil glaze

mutton cloth

white spirit

paint brush

fine grade sandpaper

paint kettle

Hints

If you cut mutton cloth it can shed tiny fibres which may spoil your work, so I prefer to pull out one of the horizontal threads which will divide the cloth without any fraying.

1 Prepare the wooden salt cellar by sanding down the surface to a smooth finish. Apply one coat of primer, then two coats of your chosen base colour. Allow to dry thoroughly and lightly sand down between each separate application.

2 Mix together equal quantities of the glaze with the coloured oil based top coat in a paint kettle. Paint evenly over the surface of the objects. With small items like salt cellars it's a good idea to stand them on a can which you can twist while painting.

3 Fold the mutton cloth to form a pad. Tuck the raw edges to the inside of the pad. Press the cloth firmly and directly onto the surface to lift off the glaze and reveal the base colour. The cloth can be rotated as it becomes saturated with the glaze.

Stippled Storage Box

Stippling is a very delicate, subtle finish which benefits from using a strongly coloured glaze over a paler base coat. It can be used on walls and large pieces of furniture as well as smaller items like this attractive storage box.

MATERIALS

coloured oil based paint – base colour

coloured oil based paint – top colour

transparent oil glaze

white spirit

paint brush

stippling brush

paint kettle

Hints

To prevent the bristles of a stippling brush from clogging with glaze, wipe frequently on a soft, clean cloth.

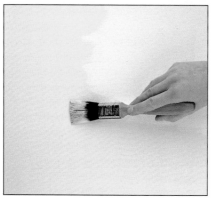

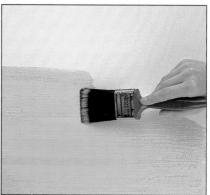

1 Paint the storage box with two good coats of your chosen base colour. Make sure the surfaces are as smooth as possible as stippling has a tendency to show every slight imperfection. Allow to dry between each coat.

2 Mix together equal quantities of the oil based top coat with the glaze in a paint kettle. If you would like to obtain a dense top colour, mix one part glaze to two parts oil colour. Apply evenly to the storage box.

3 Use a stippling brush to produce the stippled effect. Use a steady but firm tapping motion with the brush, keeping an even pressure. Avoid twisting the bristles as this will blur the effect which is a little messy and less attractive.

Limewashed Old-fashioned Food Cupboard

This is a simple technique that simulates an age-old finish. It is easy to achieve and cheaper than most techniques as it uses only a small quantity of white emulsion.

MATERIALS

white emulsion paint

5cm/2in paint brush

paint kettle

cotton cloth

masking tape

medium/coarse grade wire wool

sanding block

medium/fine sandpaper

Hints

New wood is not very porous, so you may find it useful to prepare it with a medium/coarse grade wire wool which will 'key' the surface, enabling it to hold the paint better. Always work in the direction of the wood grain.

1 The surface of the wood needs to be absolutely clean and free from dirt and grease. I used a sanding block to prepare the storage box before painting. Start with a medium grade sandpaper and finish with a fine grade for a really smooth surface.

2 Apply masking tape to those areas not to be painted. This not only protects the unpainted surface but gives a neat finish to your storage box as well. Remember to peel away carefully.

3 Dilute emulsion paint with water in a paint kettle. The amount varies according to the porosity of the wood, so test a small area first, making sure the wood grain is clearly visible. Apply the thinned emulsion a little at a time in small workable sections.

4 Use a cotton cloth to rub the still-wet emulsion into the wood. The desired effect is for the emulsion to leave a faint veil of white over the surface, with a thicker white appearance where the emulsion has held in the grain of the wood.

Dragged Doors

This technique is particularly suited for use on panelled doors as the fine dragged lines follow the panels. As doors are in constant use it's advisable to use oil based paint. An ordinary long bristled paint brush can produce satisfactory results if you don't have a dragging brush.

MATERIALS

oil based primer

white oil based paint – base coat

coloured oil based paint – top coat

transparent oil glaze

white spirit

paint brush

dragging brush or long-haired paint brush

paint kettle

sandpaper

Hints

It takes a little practice to avoid obvious brush marks at the point where the dragging starts. Use your free hand to hold the bristles firmly against the glazed surface until the brush starts to make its own mark.

1 Prepare the door by sanding down thoroughly and if necessary filling any cracks with wood filler. Paint with two layers of the base coat. Allow to dry between coats. Dilute the top coat in a paint kettle with one part paint to one part oil glaze.

2 Using the long-bristled dragging brush, apply a thin layer of coloured glaze to the door. Work with long vertical strokes to correspond with the direction of the dragging. You can paint the entire door as the glaze remains moveable for some time.

3 Start working with the door panels first. Draw the brush from top to bottom, maintaining an even pressure and keeping the brush strokes as parallel as possible. Finish by dragging the upright and horizontal bars then finally the door frame.

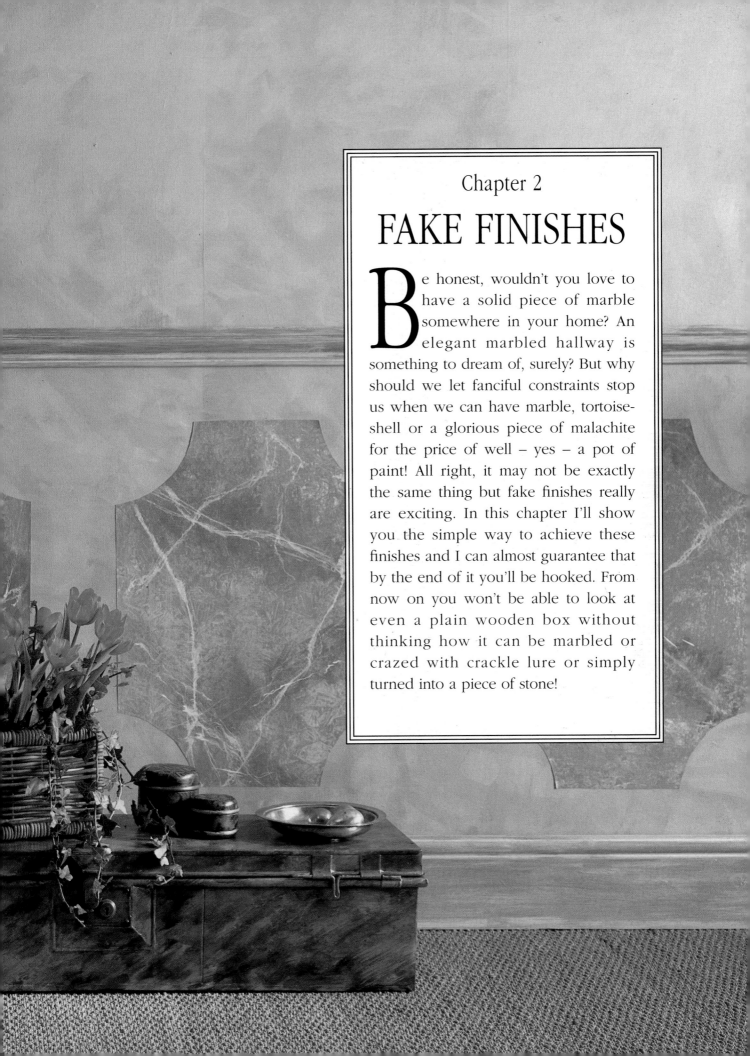

Chapter 2

FAKE FINISHES

Be honest, wouldn't you love to have a solid piece of marble somewhere in your home? An elegant marbled hallway is something to dream of, surely? But why should we let fanciful constraints stop us when we can have marble, tortoise-shell or a glorious piece of malachite for the price of well – yes – a pot of paint! All right, it may not be exactly the same thing but fake finishes really are exciting. In this chapter I'll show you the simple way to achieve these finishes and I can almost guarantee that by the end of it you'll be hooked. From now on you won't be able to look at even a plain wooden box without thinking how it can be marbled or crazed with crackle lure or simply turned into a piece of stone!

Marbled Column

There are many different types of marble and as a result there are probably as many techniques for marbling. I have broken down the basic principles to make a simplified version that always follows three basic steps: applying the glaze, distressing the glaze and veining.

MATERIALS

oil based paint – base coat

transparent oil glaze

tubes of artist's oil colour
(raw sienna and raw umber)

white spirit

5cm/2in paint brush

cotton cloth

fine-grade sandpaper

Fitch brush (or soft artist's brush)

artist's fine paint brush

Softening brush
(or long-haired paint brush)

paint kettle

Hints

It's useful to have a piece of marble to refer to when creating this effect so why not do as I did and treat yourself to a marble pastry board – or simply borrow a reference book from the library.

1 Prepare the plastic column for painting by 'keying' the surface with fine sandpaper. Apply two coats of oil based base coat. Mix a little raw sienna colour with about one tablespoon each of white spirit and oil glaze in a paint kettle and paint thinly over the column.

2 Wrap the cotton cloth into a small pad and wipe away varying sized 'pebble' shapes from the wet glaze. To get a real marble effect, rub off long thin shapes surrounded by clusters of smaller round shapes.

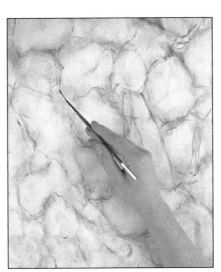

3 Use a long-haired artist's brush and lightly flick it across the whole surface. This will blur and soften the marks you have created without obliterating the colour and will still retain the overall marble effect.

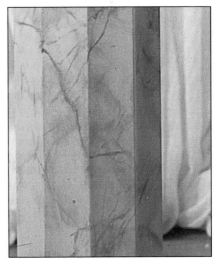

4 Thin the raw umber colour with a little white spirit and paint on the 'veins' with an artist's brush. Paint in one direction, but occasionally, cross these lines diagonally with more 'veins'. When dry, use a high gloss varnish to seal and protect the marbling.

Tortoiseshelled Trunk

Real tortoiseshell is ecologically unsound as it comes from the shell of the sea turtle, but it can easily be reproduced using oil colours.

MATERIALS

gold artist's powder

shellac varnish

transparent oil glaze

white spirit

tubes of artist's oil colours
(raw sienna and burnt umber)

5cm/2in paint brush

Fitch brush (or soft artist's brush)

softening brush (or long-haired paint brush)

paint kettle

polyurethane varnish

wire brush

sandpaper

Hints

Use some leopard-skin patterned fabric to inspire your design and choice of colours.

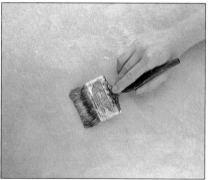

1 I 'keyed' this metal trunk by rubbing with a wire brush. If you have a wooden trunk, rub with sandpaper. Mix enough gold powder with shellac varnish to produce a creamy consistency. Paint a thin layer of colour carefully and evenly all over the trunk.

2 When this is dry, brush on a thin layer of the transparent oil glaze. Mix a little glaze individually in a saucer with each of the oil colours. Brush on the darker raw sienna in soft 'cloudy' shapes.

3 Using a soft artist's brush streak the burnt umber mix over the previous coat. Apply this in small patchy areas following a diagonal direction. At this stage the effect will resemble a leopard skin print.

4 Soften the whole area with a long haired brush using a light flicking movement while following the direction of the tortoiseshelling effect. Allow this to dry thoroughly before you continue with the next step.

5 Thin the burnt umber artist's oil paint with a little white spirit, then splatter occasional flecks over the surface. Soften this again with a long-haired brush. Leave to dry and protect the finish with polyurethane varnish.

Woodgrained Cupboard

I make no attempt to reproduce an authentic woodgrained finish here – this technique is purely for fun! I think it's quite striking and you'll be surprised how fantastically simple it is to achieve.

MATERIALS

coloured emulsion paint – base coat

white emulsion paint – top coat

emulsion glaze

5cm/2in paint brush

rubber wood graining tool

fine grade sandpaper

paint kettle

Hints

If the woodgrain isn't quite right the first time, just brush over the still wet glaze and have another attempt.

1 Prepare the cupboard for wood-graining by removing old paint, sanding down as necessary. Apply two coats of the base coat paint. Allow to dry thoroughly and lightly sand down between each application.

2 Mix equal quantities of the white emulsion with the glaze in a paint kettle. Blend well. Paint over the central panels of the cupboard as this is the only area to be woodgrained. The remainder of the cupboard is left plain.

3 Pull the rubber wood graining tool through the wet glaze. Start at the top of the panel and work slowly downwards. Rock the tool gently once or twice as you progress to achieve the dramatic woodgrain effect.

Verdigris Plaster Wall Plaque

This method of producing a verdigris finish is deceptively simple. If you use oil based colours it can be applied to any number of surfaces including bathroom walls – just use a larger brush!

MATERIALS

white emulsion paint – base coat
emerald green emulsion paint – top coat
turquoise emulsion paint – top coat
white acrylic varnish
Fitch brushes (or soft artist's brushes)
5cm/2in paint brush

Hints

If you want to give bathroom walls a verdigris finish, paint the surfaces vertically with thinned paint. This will trickle down the walls before it dries giving an interesting look to the verdigris. The varnish may also be applied in this way.

1 Paint the wall ornament with white emulsion. When dry apply the emerald green paint with a Fitch brush. Use a dabbing motion to create random splodges on the surface. It's important not to produce a solid colour.

2 While the first colour is still wet, apply the turquoise emulsion in the same manner. Occasionally overlap the colours and fill in the white areas that may remain. Using a dry brush blend the colours so they merge together.

3 When the colours are dry, brush on the white acrylic varnish, building up denser colour in some areas. Don't worry if the varnish collects in some crevices as this is all part of the effect. Leave to dry thoroughly.

Stone-blocked Walls

This is an ideal treatment to use under a dado rail, giving a solid manor house look. The blocking requires a little working out, but the effect can be most unusual.

MATERIALS

ivory coloured emulsion
paint – base coat

light, medium and dark emulsion
paint – top coats

emulsion glaze

masking tape

ruler

pencil

sea sponge

paint tray

spirit level

set square

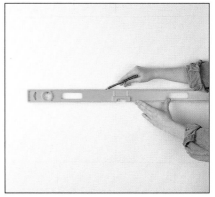

1 Apply two coats of the ivory emulsion to the wall. This will eventually show through as the 'grouting' between the fake stonework. Work out the size of your 'blocks' on paper first, then using a spirit level, set square and ruler, mark out on the wall.

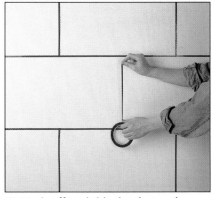

2 Mask off each block, placing the masking tape to the right of each vertical pencil mark and below each horizontal pencil mark. When eventually removed, the revealed areas will simulate mortar.

3 Your choice of coloured emulsions should ideally resemble stone colours. Mix the medium coloured emulsion with an equal part of glaze in a paint tray and sponge evenly over the stoneblocked area including over the masking tape. Allow to dry.

4 Mix the darker emulsion with an equal quantity of glaze as above, and using a clean sponge apply over the area. Overlap in irregular-shaped patches to achieve varying densities of colour. Leave to dry.

5 Mix the lightest emulsion, again with equal quantities of the glaze and apply randomly over the entire surface to soften the previous colours.

6 Allow the wall to dry completely and remove the masking tape. Peel very carefully as you may remove some of the original 'grout' colour.

Crackle Lure
Key Cupboard

Many pieces of wooden furniture benefit from a little antiquing to give them character. Using this technique, which gives a cracked paint effect, this charming little cupboard has been transformed.

MATERIALS

white emulsion primer

white emulsion paint – base coat

coloured emulsion paint – top coat

paint brush

PVA glue

cotton wool

photocopies of statues and architecture

Crackle lure kit (containing ageing oil varnish and water based varnish)

tube of artist's oil colour (umber)

fine grade sandpaper

soft cloth

Hints

To speed the drying process once the second layer of varnish has been applied, you can blow-dry the surface with a hair-dryer.

1 Prepare the cupboard by removing any existing paintwork, sanding down as necessary. Apply the white primer to the surfaces. When dry sand lightly with fine-grade sandpaper, then paint with one coat of white emulsion base coat.

2 Use a dry brush and dip the ends of the bristles into the coloured emulsion. Use a flicking motion to transfer some of the paint onto the surface. Allow to dry.

3 Use PVA to glue the photocopies within the panels of the cupboard. When dry apply the ageing varnish in a thin, even layer over the entire cupboard including the photocopies. It's a good idea to experiment first to test the drying time.

4 When the ageing varnish is almost dry, but still slightly tacky, apply a generous layer of the second, water based, varnish. Leave overnight to dry. The two types of varnish react with one another causing tiny cracks to appear in the surface of the paint.

5 To accentuate the cracks and improve the aged look, use cotton wool to rub a small amount of the umber artist's oil colour gently into the cracks. Polish off any excess with a soft cloth, taking care not to disturb or flake the cracked paint.

JOHN MAITLAND,
Duke of Lauderdale.

Malachited Decorative Wooden Pots

Malachite is a semi-precious mineral with a unique green colour. The intensity of this colour lends itself to small objects and decorative areas on walls.

MATERIALS

oil based primer

coloured eggshell paint – base coat

tube of artist's oil colour
(Monestrial green)

transparent oil glaze

white spirit

Fitch brush (or soft artist's brush)

cotton cloth

softening brush
(or long-haired paint brush)

small piece of stiff cardboard

fine-grade sandpaper

saucer

Hints

Prepare the wooden pots by rubbing to a smooth finish with fine-grade sandpaper. Apply one coat of primer. When dry apply the eggshell base coat. Allow to dry. To make the painting easier stand the pots on cans which can be turned as you work.

1 Use a fitch brush to apply the egg-shell base coat to the surface of the pots. Take care not to paint over any mouldings that may stand proud of the surface, these are better left unpainted as they will enhance the malachited pattern on the furnished pots.

2 Mix a small amount of oil glaze with the oil colour in a saucer. Use the fitch brush to apply the glaze to the surface of each pot. While the glaze is still wet, dab the surface with the cotton cloth then use the softening brush to blur the impression made by the cloth.

3 Use the torn edge of a piece of cardboard to draw the stylised circular malachite shapes into the glaze. Vary the scale of the circles by using different sizes of cardboard. Firmly press an artist's brush into the glaze and twist in a circular motion to create different formations. Allow to dry completely.

Lapis Lazuli Candlestick

Lapis lazuli is actually a blue mineral used as a gemstone which is also in the pigment of ultramarine blue. I think this treatment is such a beautiful way to transform a plain candlestick into an elegant centrepiece.

MATERIALS

tubes of artist's oil colours
(French ultramarine and Coeruleum blue)

gold paint

artist's paint brush

Fitch brush (or soft artist's brush)

Softening brush
(or long-haired paint brush)

fine-grade sandpaper

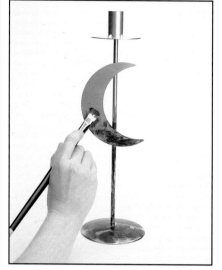

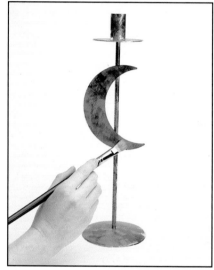

1 'Key' the surface of the metal or wooden candlestick with fine- grade sandpaper. For this treatment there is no need for a base coat. Use the Fitch brush to dab on the ultramarine oil colour in patchy areas.

2 Use the second, Coeruleum blue, oil colour, to fill in the gaps and slightly overlap the first colour. Blend the colours together with a softening brush so that there are no hard edges.

3 Dip the artist's paint brush into the gold paint and apply randomly over the surface of the candlestick to create glinting highlights.

As this candlestick is purely decorative, it is not necessary to coat with any protective varnish.

Gilded Lamp Base

Dutch Metal is considerably cheaper than real gold leaf, but looks almost the same, so I think this is a wonderfully simple way to gild. However it's a good idea to give the lamp base a final coat of varnish to stop it from tarnishing.

MATERIALS

Dutch metal leaf sheets
Japan Gold Size adhesive varnish
brown coloured varnish

Hints

Any varnish is suitable, but the brown tinted variety gives a slightly antiqued look.

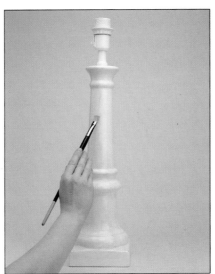

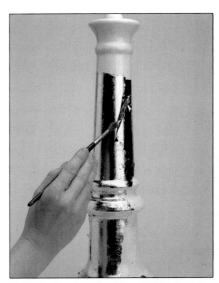

1 Gilding can be applied to any sur-face – ceramic, metal or plastic, although wooden lamp bases will need to be varnished first. Apply the Gold Size adhesive to the surface of the lamp base and wait until it is almost dry.

2 Take the sheets of Dutch Metal and press on to the tacky surface of the entire lamp base. Position each sheet carefully as you will not be able to remove them once they have come in contact with the Gold Size adhesive.

3 Rub your hand over the metal leaf to remove the loose pieces that have not adhered to the base. Make sure the surface is smooth. Paint on a layer of the coloured varnish and leave to dry.

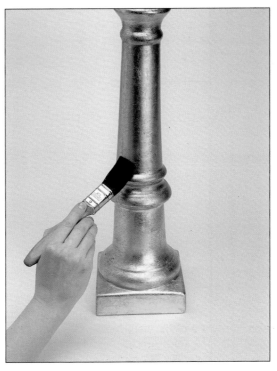

Marbled Dado Panelling

This marbling technique uses quick-drying emulsion paints. I have made the panelling from pieces of wallpaper, which means you can work on a flat surface, then simply cut and paste the marbled panel to the wall when finished.

MATERIALS

white emulsion paint – base coat	artist's fine paint brush
vinyl wallpaper	feather
2 colours emulsion paint – top coats	paint kettle
emulsion glaze	high gloss varnish
tube of artist's acrylic paint	saucer
cotton cloth	pencil
5cm/2in, and 10cm/4in paint brushes	scissors
long-haired paint brush for softening	

Hints

Give an attractive finish to your paper panels by cutting them to fit under the dado rail, then using a saucer as a guide, mark and cut out inwardly curving corners. Paste to a plain painted wall with ordinary wallpaper paste.

1 Apply white emulsion base coat to a piece of wallpaper. Mix equal amounts of the coloured emulsion with the glaze. Paint the glaze mix on to the wallpaper surface using a wide brush. While the glaze is still wet progress to the next step.

2 Twist a cotton cloth into a sausage shape, place on to the glazed surface and simply roll off the glaze as if you were rolling pastry. Shake out and retwist the cloth occasionally. Follow the same direction all the time until you have ragged the whole surface.

3 Add a little artist's acrylic paint to darken the top colour and mix with a little glaze and water. Paint sparingly in a diagonal direction from left to right, then right to left across the paper. Soften the edges by brushing lightly with a dry long-haired brush.

Plaster-effect Walls

This is one of my favourite techniques and is called plaster effect because when the three colours are combined they give a lovely soft appearance of new plasterwork.

MATERIALS

white emulsion paint – base coat

light, medium and dark emulsion paint – top coats

emulsion glaze

cellulose decorator's sponge

paint tray

paint kettles

1 Paint the walls using two base coats of white emulsion. Mix together each of the three coloured top coats with equal quantities of glaze and water. Blend each mixture well in separate paint kettles.

2 Pour a little of the darkest paint/glaze mix into a paint tray. Dip the sponge lightly into this and apply it to the surface using a circular 'scrubbing' action to create a cloudy layer of colour. This takes practice but the effect is better than using a paint brush.

4 Using a fine artist's brush dipped in the white base colour, paint diagonal 'veins' across the surface of the softened bands of colour. While these 'veins' are still wet, lightly stroke with a feather to soften their outline. When dry, finish with a high gloss varnish.

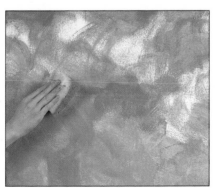

3 When the surface is dry, apply the second, medium-coloured glaze. This should be used sparingly to maintain a cloudy layer while allowing the underlying colour to show through in patches. If the second colour is too dense, rub off with a clean cloth.

4 When dry apply the lightest paint/glaze mix. Again this is applied very sparingly and is intended simply to soften the whole effect while giving just a hint of colour. If at this stage you feel the overall effect is too light, reapply the first colour.

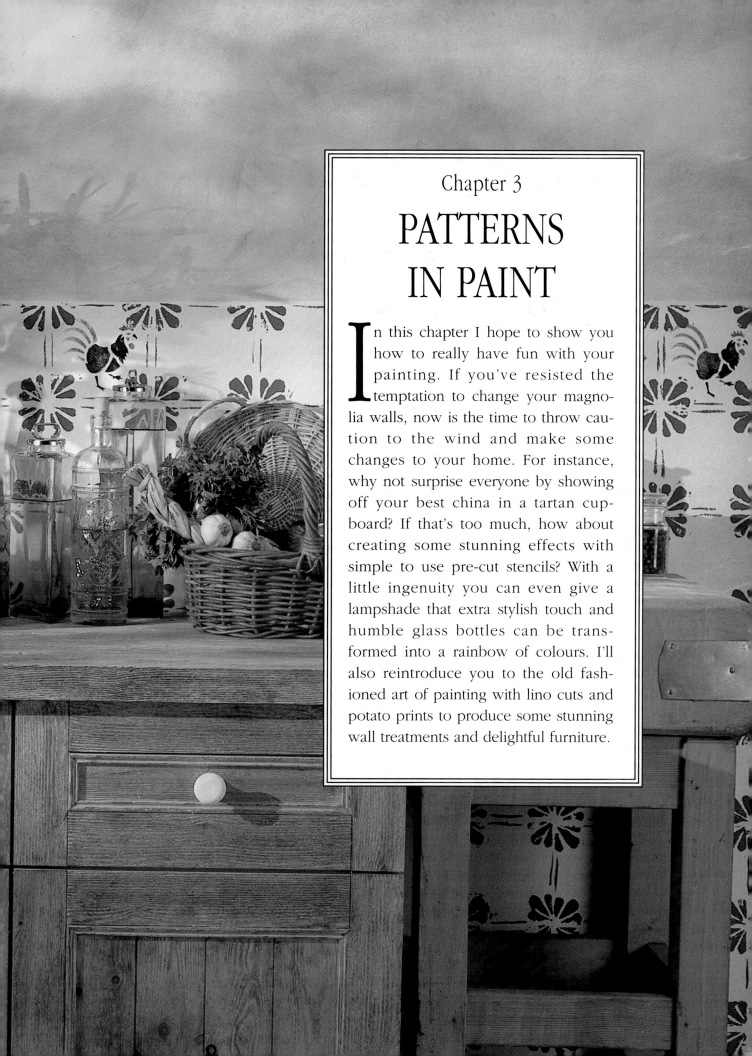

Chapter 3

PATTERNS
IN PAINT

In this chapter I hope to show you how to really have fun with your painting. If you've resisted the temptation to change your magnolia walls, now is the time to throw caution to the wind and make some changes to your home. For instance, why not surprise everyone by showing off your best china in a tartan cupboard? If that's too much, how about creating some stunning effects with simple to use pre-cut stencils? With a little ingenuity you can even give a lampshade that extra stylish touch and humble glass bottles can be transformed into a rainbow of colours. I'll also reintroduce you to the old fashioned art of painting with lino cuts and potato prints to produce some stunning wall treatments and delightful furniture.

Stencilled Chest of Drawers

Even the cheapest chest of drawers can become something special when you pattern it with a stencilled design. As there are so many pre-cut stencils, you'll be spoilt for choice, but I particularly like this rope and tassel design.

MATERIALS

white acrylic primer

2 colours emulsion paint – top coats

pre-cut stencil

stencil brush

spray adhesive

paint for stencil

paint brush

saucer

Hints

You can achieve a more subtle effect, when stencilling, if you use your paint sparingly and blend a slightly lighter or darker shade over the first colour, as I did with the chest of drawers.

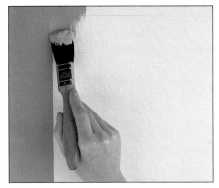

1 Prepare the chest of drawers with acrylic primer. If you are working with a new, unpainted piece of furniture as I did, you won't need a base coat. Older items of furniture may need more attention. Paint the framework of the chest with emulsion paint.

2 As an interesting variation I thought it would be fun to paint a panel on the drawers in a toning shade of emulsion paint. I used pale to mid tones for a more sophisticated look, but brighter, primary colours are probably better for childrens' rooms.

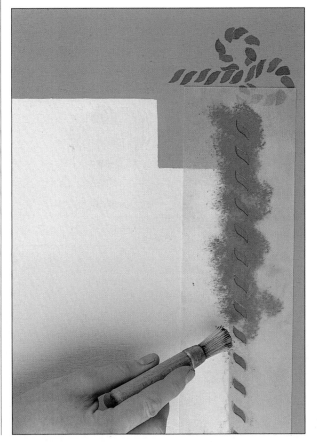

3 Position the stencil on the drawer area with spray adhesive. Put a little stencil paint onto a saucer. Dip the stencil brush into the paint and using a circular motion carefully apply the paint to the cut area of the stencil.

Stencilled Lampshade

I have used a sharp blade to cut along a few of the stencilled outlines, which gives this pretty lampshade a delicate three-dimensional pattern and allows the light to pass through.

MATERIALS

thick cartridge paper

craft knife

pre-cut stencil

acrylic stencil paints

stencil brush

saucer

adhesive

Hints

If you don't have a stencil brush use any small square ended brush or trim the bristles of an ordinary paint brush.

1 Plain lampshades are perfect to stencil but if you want to reproduce an existing shade simply remove it from its frame and draw around it on a piece of cartridge paper. Cut out the flat shape, allowing 2.5cm/1in extra all round for turnings.

2 Position your stencil on the shade with spray adhesive. Pour a little stencil paint into a saucer and apply the paint sparingly with light circular strokes to the shade. Remember if you are stencilling a fabric-coated shade, use fabric stencil paints.

3 When the paint is dry, use a craft knife to cut along some of the stencilled outlines and ease away from the shade so a little lamp light shines through. Snip the turning allocation of a new lamp shade, fold these edges under the frame and glue with strong adhesive.

Stencilled Floor Cloth

Floor cloths are as old as time but they never seem to go out of fashion. I chose this pretty flower stencil for my cloth to use in a bedroom or conservatory. Remember when the cloth has been sealed, it is extremely hard wearing.

MATERIALS

canvas – floor cloth weight

pre-cut stencil

emulsion paints for stencilling

stencil brush

PVA glue

paint brush

polyurethane varnish

scissors

masking tape

Hints

Roll and press the stencil brush firmly into the canvas as you are painting to make sure that the emulsion paint penetrates the weave.

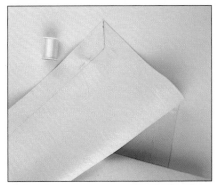

1 Cut the canvas according to the size of floor cloth you want. This canvas is available in very large widths so you could get two small cloths from one width. Turn under the raw edges of the canvas and stitch on a machine or glue turnings in place.

2 Position the stencil to form a border on the cloth. Hold in place with masking tape. To produce depth and subtlety in my design, I used three paint colours. Before each shade was dry, I carefully blended the next shade on top while allowing the first to show through.

3 Use the same technique to stencil in the central design. Before doing so it's advisable to determine the middle of your cloth. The simplest way to do this is to fold it into quarters, pressing firmly along the creases to pinpoint the exact centre.

4 When the emulsion is quite dry, paint on a layer of PVA glue. Leave this to dry. Repeat the process until you have built up four layers of adhesive. When the last coat is dry, protect the cloth with a final layer of polyurethane varnish.

Stencilled Larder Cupboard

I love the simplicity of Shaker furniture so the inspiration for this larder cupboard comes from that very unique style of furniture.

MATERIALS

white acrylic primer

coloured emulsion paint – base coat

coloured emulsion paint for stencil

cartridge paper

craft knife

stencil brush

sandpaper

pencil

ruler

masking tape

newspapers

Hints

Use up any left over emulsion paints for stencilling or buy inexpensive sample pots of colour. Alternatively look at the ranges of stencil paints.

1 Prepare the cupboard for stencilling by priming, then apply two layers of emulsion base coat. Rub with sandpaper after applying each coat to produce a perfectly even surface.

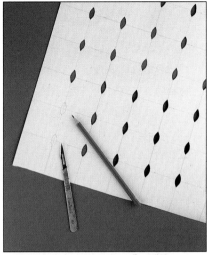

2 Cut cartridge paper to fit the cupboard door panel. Using a pencil and ruler divide the paper into 5cm/2in squares. At the intersection of each square draw a leaf shape. Cut out with a craft knife.

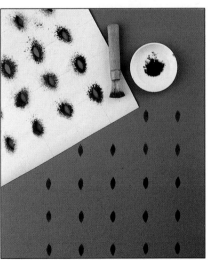

3 Stick the paper stencil to the panel with masking tape and paint through your cut-out stencil shape using the coloured emulsion paint. Remove the masking tape carefully to prevent pulling away any of the base coat.

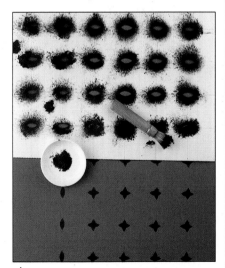

4 Turn the stencil paper through 90 degrees. Reposition on the door panel with masking tape so the cut-out leaf shape now crosses through the stencilled area. Again stencil through the existing leaf shape with coloured emulsion paint.

Vegetable Printed Kitchen Table

Who would believe that the humble potato print we all experimented with at school could produce such an ingenious pattern for a table top? I think it gives the table a pretty, French provincial look.

MATERIALS

white acrylic primer

white emulsion paint – base coat

coloured emulsion paint – printing

potato

knife

sandpaper

felt tipped pen

pencil

ruler

saucer

newspaper

Hints

As you work with the potato it can become very slippery, so you could make a handle for it by sticking a fork into its back. Alternatively wear rubber gloves to get a better grip.

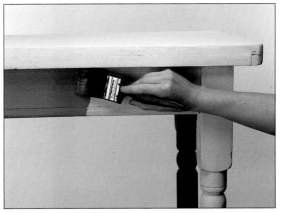

1 Prepare the table by first sanding down then priming the top. Paint the legs with basecoat only. Don't worry if the table top is uneven as this will add to its country look. Use the pencil and ruler to divide the top into 15cm/6in squares.

2 Cut a large potato in half and draw your design onto the surface with a felt tipped pen. Cut around the pattern shapes in the potato and remove the excess pieces so the pattern stands proud. I cut two shapes, one to outline the tile, shown here, the other to print the design.

3 Pour a little of the emulsion paint into a saucer and dip the potato into the paint. Test the pattern on a sheet of newspaper before printing. I printed the tile outline over the pencil drawn lines first, then applied the second design to the centre.

Tartan Tallboy

Tartan is popular so be adventurous and paint this striking design over your dullest pieces of furniture – it's bound to make a sensational transformation!

MATERIALS

white acrylic primer

coloured emulsion paint – base coat

5 coloured emulsion paints – for tartan effect

fine artist's brushes

pencil

masking tape

sandpaper

ruler

Preparation

This is a good paint technique to use on a battered old tallboy as the patterning conceals any defects. Prepare old drawers by sanding thoroughly, then apply the acrylic primer. Paint with two coats of the coloured emulsion base coat.

Hints

You will find it easier to paint your tallboy in stages. Begin with the frame and then the drawers. If the prospect is too daunting, paint the drawers only.

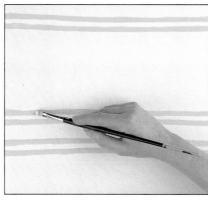

1 Decide where you want your tartan pattern to be painted and mark out the first lines using a pencil and ruler. The first colour is applied as parallel lines with a 2.5cm/1in gap between and a 15cm/6in space between the sets of parallel lines.

2 Apply the second colour, again in parallel lines, but this time paint them vertically so they cross through the previously painted lines. You may now feel more confident about painting free hand, but I don't think it really matters if the lines are wobbly.

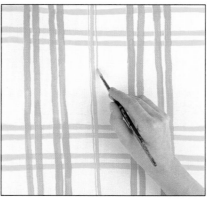

3 Apply the third colour, again vertically, in the middle of the 15cm/6in gap. I drew one thick line with a fine line on each side, but choose whatever style of lines you prefer.

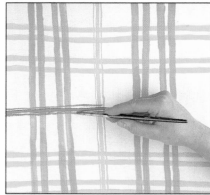

4 Use a slightly darker shade of the previous colour to reproduce a similar trio of horizontal lines. Again these should be painted within the 15cm/6in gap left in step 1.

Tartan Door Knobs

This really is a clever way of adding decoration to your drawers or cupboards, especially if you can't face the all-over tartan effect.

MATERIALS

white acrylic primer

coloured emulsion paint – base coat

coloured emulsion paints – for tartan

fine artist's brushes

polyurethane varnish

Hints

Knobs are always tricky to paint on a flat surface so you may find it easier to insert a long screw into the base of each to provide a handle. Support on a bottle while drying.

1 Prepare your knobs by priming them, then apply one coat of base coat. This base coat colour can either match or contrast with the colour of the drawers or cupboard for which the knobs are intended.

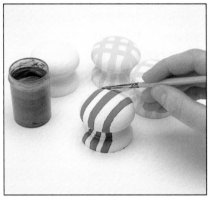

2 With small objects it really isn't necessary to use masking tape as a guide but it may be better to plan the tartan lines in pencil on your first knob. Paint on thick lines horizontally and vertically to form a chessboard effect.

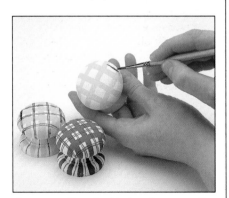

3 Paint a fine double line horizontally and vertically through the centre of the chequered pattern to produce a simple tartan effect. Leave to dry and finish with one layer of varnish.

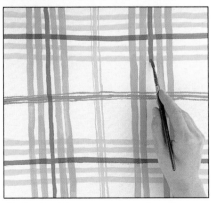

5 The final lines are painted in the boldest colour of all and are positioned within the 2.5cm/1in gap of the original vertical and horizontal lines.

Striped
and Freehand
Painted
Walls

This is an unusual method of painting on walls which not only gives you a classic pattern to produce in any colour but which also gives a plain room instant character.

MATERIALS

white emulsion paint

2 colours emulsion paint – top coat

10cm/4in paint brush

artist's paint brush

dark brown emulsion paint – for column

plumb-line

pencil

paint kettle

cotton cloth

Hints

Stripes can be masked off with low-tack masking tape if you are working with furniture or small areas.

1 Use the plumb-line to plot where you wish your stripes to appear. Mark with a pencil. Our stripes were approximately 45cm/18in wide. It's advisable not to make your stripes too narrow as this will make the overall effect too busy.

2 With your first colour of emulsion paint, infill one of the broad stripes using the wide paint brush. Apply swiftly with vertical strokes. It doesn't really matter if the colour is uneven. Repeat this shade for every other stripe.

3 Use the second colour to paint between the existing stripes. I chose a toning shade as a contrasting colour would be too busy and I wanted the finished striped effect to look rather soft and subtle.

4 Mix one part of the white emulsion paint to four parts of water in a paint kettle. Use a cotton cloth to wipe this mix all over the striped wall using broad vertical strokes to soften the look and provide a powdery feel.

5 Paint in a simplified freehand column within a painted stripe at various points on the wall. The top detail should be painted just below the cornice and a similar bottom detail just above the skirting board. The two shapes are linked with the column outline.

Stylised Painted Cupboard

This cupboard has a regular pattern that is remarkably easy to paint. You could enlarge the scale and paint the same design over your walls as a brighter alternative to paper.

MATERIALS

white acrylic primer

coloured emulsion paint – base coat

3 colours emulsion paint – for pattern

fine artist's brush

paint brush

tracing paper

thin cardboard

pencil

Hints

Any stylised leaf shape would look good in this pattern. Experiment with a few alternatives before you commit yourself.

1 Prime and base coat the cupboard. You'll find it easier to paint furniture with legs if you turn the piece upside down and paint the legs first before painting the rest.

2 If you want to copy my simple leaf design trace the outline from this page. Use this to make a cardboard template and draw at random all over your cupboard. Infill this outline with boldly coloured emulsion paint.

3 Trace the central oak leaf shape again from this page and make a second card template. Position within the original leaf shape and draw around. Infill with a lighter shade of emulsion paint.

4 The final fanciful detail on my cupboard uses a contrasting colour and is a simple curved outline created with a quick sweeping freehand movement of the brush.

Marquetry
Floorboards

As a border on bare wooden
floors this method looks
stunning and belies the true
simplicity of the technique.

MATERIALS

3 colours wood varnish

stencil brushes

pre-cut stencil

masking tape

clear gloss varnish

white spirit

Hints

Check the floorboards for irregulari-
ties. Secure loose boards and sink
all exposed nail heads using a ham-
mer and nail punch. If necessary
add artist's oil colours to the varnish
to achieve a required shade.

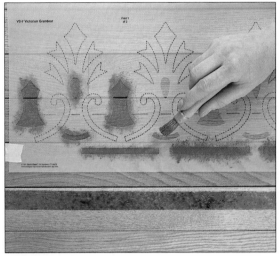

1 Create a border for the stencil design by placing two strips of masking tape on the floor about 5cm/2in apart. Infill the stripe between with the darkest shade of varnish. When dry carefully peel away the masking tape.

2 Choose a stencil that uses at least two different colours. Make sure the stencil sheets are solvent resistant. Position the first stencil sheet against the bottom edge of the varnished stripe and paint through the stencil with coloured varnish onto the floorboards.

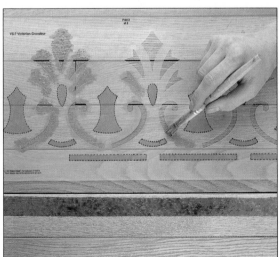

3 When the varnish is absolutely dry, position the second layer of your stencil and apply the next varnish colour. Always hold the stencil in place with masking tape and ensure that the registration lines match accurately.

4 Wait 24 hours to allow the varnishes to dry thoroughly then coat the whole design with one or two coats of hard surface varnish.

Glass-painted Storage Bottles

These glass bottles look particularly attractive in a bathroom setting and can be used to hold bath oils or lotions.

MATERIALS

specialist glass paint colours

paint tray

wooden skewer

plastic bag

Hints

Two or more glass paint colours can be swirled together in the water to create an interesting pattern.

1 Fill a paint tray with water and dribble in a few drops of the glass paint. These will float on the surface so you will need to swirl them around thoroughly with a skewer. The colour does not blend into the water but remains as swirls in the water.

2 Briefly dip the glass bottle into the paint tray to coat one side with the solution. Remove and rotate the bottle before dipping again in the solution to coat the other side. Any paint on your fingers may be easily cleaned with white spirit.

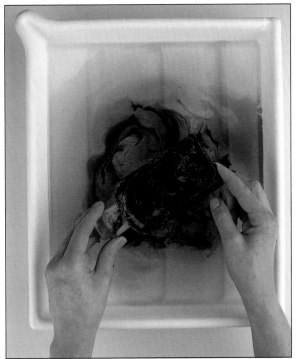

3 Stand on a plastic bag to dry. The colours won't run. Simply peel the bag away from the base of the bottles once they are dry.

Lino Printed Wall

I love 'cheating' designs and this is one of the best ways to produce a Delft tile look for a kitchen or bathroom wall without the expense of buying the real thing!

MATERIALS

15cm/6in square of lino

lino cutting tool or heavy duty craft knife

coloured emulsion paint

artist's acrylic paint

pre-cut stencil

artist's brush

felt tipped pen

newspaper

Hints

Delft tiles can be bought from tiling specialists and feature many pretty designs. You could consider buying just one tile to use as inspiration for your own stencil.

1 Draw a simple pattern in each corner of the lino square with a felt tipped pen. Carefully cut around the design with a lino cutting tool. Remove the excess lino so that the pattern stands proud ready for printing.

2 Using an artist's brush apply the emulsion paint to the patterned areas on the lino tile. Press the tile onto newspaper until the pattern shows clearly, then begin patterning your chosen area of wall. The pattern should produce a square tile effect.

3 Use a fine artist's brush to paint the coloured emulsion paint onto the wall. This should simulate the fine grout lines that exist between real tiles. These lines can be painted in sketchily to enhance the design.

4 Take a small pre-cut stencil design – I chose a little chicken – and use to stencil, here and there, in the middle of your previously stencilled tile area. Use the same coloured emulsion paint, but lowlight with a darker shade of artist's acrylic paint.

Chapter 4

PAINTED INSPIRATIONS

This is the kind of painting I really enjoy. By now you may have turned your hand to some of the projects in the earlier chapters and feel encouraged to embark on something a bit more challenging. Well, now you can indulge in a few 'tricks of the eye' with designs that will fool all your friends. These designs take a little more planning and require more free interpretation, but I've also included some ideas using photostats from books, which are a wonderful way of cheating clever designs.* And I'm sure you will be able to try simple ideas like my decorative candlestick and vases. Hopefully by the end of this chapter you'll also be inspired to create fanciful projects of your own.

* Photostats from books should only be used for personal projects not for items to be sold.

Trompe l'oeil Card Table

You'll have to look twice at this table to realise that the cards and game board are actually painted! But don't be put off, you'll be surprised how simple trompe l'oeil is to achieve and it cannot fail to make an impression.

MATERIALS

white acrylic primer

coloured emulsion paint – base coat

coloured emulsion paint – for effect

paint brush

artist's brush

polyurethane varnish

cards

game board

pencil

low tack masking tape

tracing paper

sandpaper

Hints

Any board game or cards can be used for inspiration: backgammon, chess or monopoly all work well.

1 Prepare the card table for painting. Sand down then apply the acrylic primer. When dry position the cards, game board and pieces. Draw around the shapes with a soft pencil. You don't need to use all the cards or the whole game board.

2 At this stage you may find it easier to mask the inside edges of the card and board shapes with low tack masking tape. Apply the base coat to the whole table except for the areas of trompe l'oeil. Use an artist's brush to paint around smaller outlines.

3 Paint the inside areas of the trompe l'oeil designs with cream emulsion paint using an artist's paint brush. Take care not to overlap the base coat by using masking tape if you don't have a steady hand.

4 Trace the card and board patterns. Reverse the tracing paper and go over with a soft pencil. Trace the outline right side up on to the playing cards and boardgame. Fill in the design with paints using a fine artist's paint brush. When dry coat with polyurethane varnish.

Decorative Painted Candlestick

So often you can find inexpensive plain, black metal ornaments which with a simple paint treatment can be totally transformed into something much more colourful and interesting.

MATERIALS

coloured emulsion paint

white emulsion paint

gloss paint

paint brush

artist's paint brush

paint kettle

1 My candlestick was embellished with a twirling leaf design, so it was a natural inclination to highlight these areas with leaf green emulsion paint.

2 Mix equal quantities of white emulsion paint with water in a paint kettle. Apply lightly to the painted areas to give a soft, powdery appearance to suggest age.

3 Paint in other details like flowers with gloss paint to create contrast and an interesting high shine detail.

Freehand Painted Server Cabinet

If you don't have any confidence with freehand artwork, just try this stylised New England tree design. It really is so simple if you follow my steps.

MATERIALS

oil based primer

2 colours oil based paint – top coat

artist's oil colours

artist's brush

white spirit

sandpaper

antiquing wax

cotton cloth

Hints

Any cupboard door with a central panel or even a panelled internal door can be painted in this way.

Prepare the cabinet by first sanding down, then apply the oil based primer. Using the darkest oil based top coat, paint the cupboard surround. The door panels should be painted with the lighter oil based top coat.

1 I have broken down the elements for the simple tree. First paint the tree trunk and branches using the artist's oil colour and brush. Use long sweeping strokes for the branches.

2 Add the simple leaf shapes to the branches by pressing the bristles of the brush firmly against the surface and removing carefully. The leaves don't need to touch the branches.

3 Paint in the simplified bird using three shapes to suggest its head, body and tail. Add the small detail of the plant at the base of the tree in the same way as you painted to the leaves.

4 Apply a little of the antiquing wax to a soft cloth and rub gently all over the surface of the cupboard. This will give an attractive aged look to the surface of the furniture.

Photocopy-
patterned
Blanket Box

This may look like a very
complicated pattern but it's
actually created with cut
pieces of a photocopied
design.

MATERIALS

white acrylic primer

coloured emulsion paint – base coat

emulsion glaze

acrylic artist's paint – ochre

photocopied design for border

PVA glue

coloured antiquing wax

cotton cloth

saucer

10cm/4in paint brush

masking tape

scissors

Hints

A stencilled border would work just
as well as the photocopies if you
can't find a suitable design.
Prepare the blanket box by applying
the primer and the coloured base
coat. It's a good idea to paint both
the inside and outside of the box.

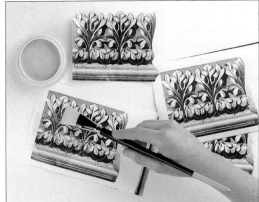

1 Make photocopies of your chosen design. Put a squeeze of ochre artist's acrylic paint in a saucer and mix with emulsion glaze. Paint over each photocopy with the glaze to give them a faded appearance.

2 When dry cut out the required design from the photocopies and glue to the blanket box with PVA. I arranged the photocopies as a border but you could use a central motif if you prefer.

3 Put some antiquing wax on to a soft cotton cloth and carefully apply it to the blanket box. Be particularly careful when covering the photocopies. Use a circular motion to achieve a soft, cloudy effect.

Reverse-stencilled Dressing Table

Transforming an ordinary pine dressing table into a work of art is so simple with this technique. I think this would also work particularly well using strong primary colours.

MATERIALS

white acrylic primer

white emulsion paint – base coat

coloured emulsion paint – top coat

artist's acrylic paint

scissors

plain paper

paint brush

set of playing cards

spray adhesive

masking tape

Hints

Instead of the card motifs use simple flower or leaf shapes. Make sure the paint is completely dry before you carefully remove the masking tape.

1 Apply the primer, then the base coat to the dressing table. Copy the hearts, clubs, diamonds and spade shapes from playing cards onto plain white paper and cut out.

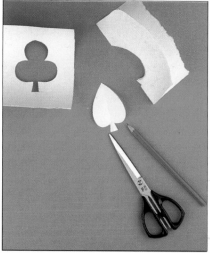

2 Position the cut-out shapes on the drawers and sides of the dressing table with spray adhesive. Mark a 15cm/6in square around each shape with masking tape.

3 Fill in the square around the shapes with black artist's acrylic paint and leave to dry.

Antiqued Wardrobe

This is a very distinctive technique to give an ordinary wooden wardrobe an aged look that could fool generations to come.

MATERIALS

white acrylic primer

coloured emulsion paint – base coat

crackle glaze

coloured emulsion paint – top coat

sanding block

sandpaper

wire wool

artist's acrylic paint – burnt umber

paint brush

paint kettle

Hints

The colours of the base and top coats should always contrast to give a strong final effect.

1 Prepare the wardrobe by first applying the primer and then the base coat. It's a good idea to sand down between coats to achieve a smooth finish. If the wardrobe is secondhand you may need to vacuum the inside.

2 When the paint is dry, apply a thin layer of crackle glaze to the wardrobe. I decided to crackle glaze just the panels to emphasise their shape and give them a distinctive look. Allow to dry.

3 Apply the coloured emulsion paint top coat to the remainder of the wardrobe, then finally to the panels. Work quickly over the panels. The paint will interact with the glaze and start to crack very quickly.

4 When the emulsion top coat on the remainder of the wardrobe is dry, use the sanding block to rub away small areas to give the effect of wear. I decided to sand areas of the centre panel as well, but this is not essential.

5 Dilute the raw umber artist's paint with water in a small paint kettle. Dip the wire wool into the paint and rub over the wardrobe to remove further patches of emulsion. This leaves an attractive overall tinted colour.

Transfer Circular Table

I have taken this motif from one of the many design source books available in bookshops and libraries.

MATERIALS

white acrylic primer	transfer paper
emulsion paint – base coat	tracing paper
artist's acrylic paint	pencil
white emulsion paint	paint brush
emulsion glaze	paint kettle
design source book	

Preparation

Prepare the table for patterning by first priming, then applying the base coat. For the best finish it's a good idea to prime and paint the underside of the table first, then the top.

1 Photocopy your design from the source book and enlarge to the required size. Trace the design using a soft pencil.

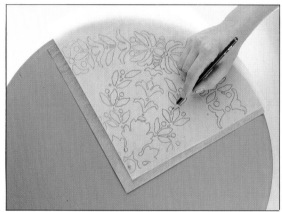

2 Place the transfer paper on the table, cover with the tracing paper, then firmly outline the design so that it transfers to the table.

3 With your chosen artist's colour paint in the design. Leave to dry. I've chosen to use only a single colour for a dramatic effect, but several colours would work just as well.

4 Mix equal quantities of the white emulsion paint with the glaze in a paint kettle. Blend well. Paint a thin coat over the table top to seal your design and to give it a faded appearance.

Colour-rubbed Pedestal

MATERIALS

coloured eggshell paint

transparent oil glaze

cotton cloth

wire brush

Fitch brush (or soft artist's brush)

paint kettle

paint brush

1 Prepare the pedestal by rubbing all over with a wire brush. Mix equal quantities of the eggshell paint with the glaze in a paint kettle. Paint liberally over the pedestal.

2 While the glaze is still wet, rub the surface with a cotton cloth to remove most of it, so that only the recessed areas retain the glaze.

Pen and Ink Clock Table

A beautiful antique fob watch was the inspiration behind the unusual clock face on this circular coffee table.

MATERIALS

white oil based primer

black waterproof ink

ink pen and nib

string

ruler

pencil

crackle lure kit (antiquing varnish and water based varnish)

artist's oil colour – burnt umber

cotton cloth

polyurethane varnish

paint brush

Preparation

Prime the table and allow to dry. Mark the circular clock face on the surface of the table. To do this, tie a length of string to a pencil, hold the end in the centre of the table and pull taut. Draw the circle about 2.5cm/1in from the edge of the table. Add another circle 2.5cm/1in inside this. Use the black ink to define the outline.

Hints

I've used Roman numerals for this table but if you wanted to paint a table for a child's room, copy the face of a modern analogue watch.

1 Fill in the details of the clock face using a pencil and ruler to outline the Roman numerals. Infill with the ink pen using the waterproof ink. If you don't have the confidence to do this freehand, trace the numerals from a reference book and using transfer paper reproduce onto the table top.

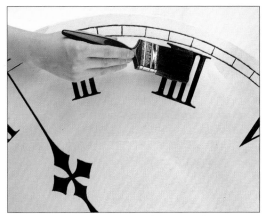

2 Apply the antiquing varnish to the surface of the table with a paint brush. Leave to dry until slightly tacky. Apply the water based varnish. Leave to dry. During the drying process fine hairline cracks will appear on the surface.

3 Take a little of the raw umber artist's colour and using a soft cloth rub into the surface to define the cracks in the varnish. This process enhances the antiqued look of the face. When absolutely dry, seal with a coat of polyurethane varnish.

Photocopied Cherub Bedside Cabinets

For the true romantics this is a lovely way to decorate plain bedside cabinets. The photocopies form the outline for the painted cherubs but the effect looks like an original hand drawing.

MATERIALS

white acrylic primer

3 colours of emulsion paint – base coat

artist's acrylic paints

artist's brush

photocopies of cherubs

PVA glue

paint kettle

masking tape

scissors

Preparation

Prime the bedside cabinets and allow to dry. Mix equal quantities of one emulsion paint base coat with water in a paint kettle and apply to the body of each cabinet.

Hints

If you are decorating a child's room, painted photocopies of animals, clowns or their favourite heroes would be an inspired alternative to stencils. Don't be afraid to use a strong shade when colourwashing because the final result will be much paler.

1 Mark a 2.5cm/1in border around the edge of each cabinet door using masking tape. Mix equal quantities of the second emulsion base coat with water and apply to the central section within the masked area. You can use either a lighter variation of the first colour or a contrasting shade.

2 Remove the masking tape and reposition to protect the painted panel area. Paint the revealed border with the third emulsion paint colour, again mixed with an equal quantity of water. When removing masking tape avoid peeling away any layers of paint.

3 Cut out the cherub shapes from the photocopies and using the PVA glue position on the bedside cabinets. My cherubs were enlarged to fit neatly within the central panel of the cabinet door fronts.

4 Use the acrylic paints to colour the cherubs. At this stage you can let your imagination go wild with the colours. Or if you prefer, just leave the cherubs perfectly plain.

Freehand Painted Tray

Any metal or unglazed ceramic can be transformed in this way. I used a classic pattern to add an elegant touch to this particular tray.

MATERIALS

coloured emulsion paint - base coat

Japan Gold size

Dutch Metal leaf

pencil

artist's paint brush

polyurethane varnish

Hints

This technique can be applied to many small objects, including the vase that's featured on our front cover. To embellish the patterning you can add swirls of colour and star shapes which I painted on freehand.

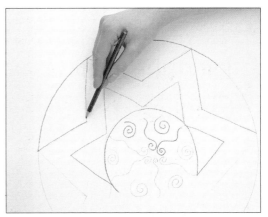

1 Mark your chosen design with a pencil. I prefer simple patterns so this star shape with a circle in the middle looked elegant without being too overpowering and can be easily copied.

2 Use the coloured emulsion to decorate the pattern. Use only two or three colours for the most impact. Don't be afraid to choose bold colours. Paint Japan Gold size adhesive on to the tray where the Dutch Metal leaf is to be applied. I decided to use it on the star outline and spirals.

3 When the adhesive is just tacky, press the Dutch Metal leaf onto the surface and peel off its backing paper. Allow to dry and coat with one layer of varnish to seal.

Lacquer and Crackle Lure Serving Tray

This is a quick and simple way of faking what is traditionally a very time-consuming paint technique.

MATERIALS

white oil based primer

red gloss paint

Crackle Lure kit (antiquing varnish and water based varnish)

artist's oil colour – raw umber

cartridge paper

plain paper

wet and dry sandpaper

cotton cloth

ruler

scissors

pencil

paint brush

masking tape

polyurethane varnish

Hints

Always use varnish in a dust-free environment. Even tiny specks of dust will spoil the surface of your tray.

1 Prepare both surfaces of the wooden serving tray by rubbing with sandpaper and then applying a coat of primer. When completely dry apply a coat of the red gloss paint.

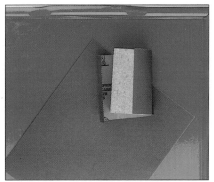

2 Use wet and dry sandpaper to "key" the gloss painted surface. Paint on a further coat of red gloss paint. Repeat this process again to build up three layers of gloss paint.

3 Tape cartridge paper to a flat surface and coat with the oil based primer. When dry apply the crackle lure antiquing varnish. When this is just tacky, paint on the water based varnish. As this dries fine cracks will appear on the surface.

4 Put a little raw umber artist's oil paint on a soft cloth and rub into the cracked surface on the cartridge paper. This will define and highlight the cracks for a more aged appearance.

5 Make a template of a star shape on plain paper, then use as an outline on the prepared cartridge paper. Cut out. Coat the tray with polyurethane varnish and while it is still wet position the paper star. When the varnish is dry seal with a further coat of varnish.

Trompe l'oeil Venetian Villa Drawers

This clever design was inspired by the brilliant style of marquetry that was popular at the turn of the century.

MATERIALS

white acrylic primer

white emulsion – base coat

coloured emulsion paint – top coat

emulsion glaze

acrylic artists' paints

low tack masking tape

ruler

pencil

artists' brushes

paint kettle

paint brush

sandpaper

Hints

If a Venetian villa isn't quite your style, look at pictures of alternative architectural designs for inspiration.

1 Remove any knobs or handles from your set of drawers. Prepare the surface for painting by applying the primer and one coat of the base coat. Sand down between each coat.

2 Mix equal quantities of glaze with the coloured emulsion top coat in a paint kettle. Add a little dark brown artist's acrylic paint to give depth to the terracotta colour. Apply to the surface with random circular movements to give a cloudy effect.

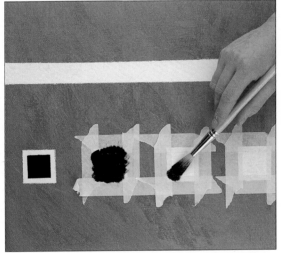

3 Using a pencil and ruler copy the positions of the windows and doors on the front of the drawers. Apply masking tape to the outside edges of these pencil marks. Paint the areas within the tape using cream artist's paint. When dry remove masking tape very carefully.

4 Use low tack masking tape to define the smaller areas within the window and door shapes. Paint with brown artist's paint inside these masked areas. Leave to dry, then remove the tape. For real authenticity add doors and windows to the sides of the chest.

Dulux paints market a comprehensive and diverse range of paint both in their ready-to-use and Definitions ranges. Throughout Simply Paint we have used the following colours from the Definitions range.

Sponged walls: Gambit (2050Y90R); Chintz (1030Y70R); New Inspirations Ready to Use Range–Wild Sage; Charades (0015Y70R)
Sponged coffee table: Appleton (2020G); Dewberry (1015G); Pale Face (0710R)
Sponged ceramics: Specialist ceramic paints
Colour-rubbed linen basket: Album (2020R50B)
Colour-rubbed bathroom storage rack: Fresh Winds (2030B50G)
Combed picture frame: Summer Wish (1040B50G); Ecstasy (1050Y90R)
Dragged doors: Gondola (5040R80B)
Colour-washed panelling: Schooner (3050R80B); Double Dutch (1040R80B)
Ragged chair: Monsoon (1030B50G)
Mutton-clothed salt cellar: Velvet Moon (6040R70B)
Ragged chair: Tea Dance (2040Y80R)
Limewashed old-fashioned food cupboard: White emulsion paint
Marbled panelling: Roller Coaster (2040R)
Woodgrained cupboard: Corniche (1050R80B)
Verdigris wall plaque: Knot Garden (4040B90G); Bath Salts (1040B90G)
Stone-blocked wall: Tartare (0010Y); Flannel (1505Y10R); Feather (3010Y10R)
Plaster-effect walls: Brique (4040Y80R); Tea Dance (2040Y80R); Lotus Shell (1015Y60R)
Crackle lure key cupboard: New Inspirations/Ready to Use Range–Wild Sage
Malachited wooden pots: Park Drive (2070G)
Stencilled chest of drawers: Soft Light (1010R20B); Velvet Jacket (2030R70B)

Stencilled floor cloth: Velvet Jacket (2030R70B); Painted Lady (1090Y80R); Solstice (0070Y10R); Puppet (1070G); Tree Top (4050G20Y)
Stencilled larder cupboard: Tarn (5020B30G)
Lino printed wall: Schooner (3050R80B)
Vegetable printed table: Silk Ripple (2050R80B)
Tartan tallboy: Tartare (0010Y); Beacon (0060Y10R); Fresh Winds (2030B50G); Prospero (1020R70B); Musselbed (5010R50B); Jester (0090Y70R)
Tartan door knobs: April Sky (2040R80B); Painted Lady (1090Y80R)
Striped walls: Wheatlight (1015Y); Pyramid (1030Y20R); Peppermill (8010Y10R)
Stylised cupboard: Hubblebubble (1030R70B); Painted Lady (1090Y80R); Dawn Surprise (0050Y); Slipper (3040B)
Trompe l'oeil card table: Excursion (6020G10Y); Backgammon, Scallywag (2020Y20R); Festival (1090Y90R)
Painted server cabinet: Wheatlight (1015Y); Sea Rhythm (4040B)
Photocopy-patterned blanket box: Helter Skelter (0030Y)
Reverse-stencilled dressing table: Winter Warmer (2070R10B)
Antiqued wardrobe: Double Dutch (1040R80B); Tartare (0010Y)
Pedestal table: Parterre (3030B90G)
Cherub bedside cabinet: Racing Silk (0740B90G); Spring Fayre (1030Y10R); Birthday Girl (0040R10B)
Painted tray: Equinox (3040R60B); Apothecary (3070R70B); Buff Beauty (2030Y30R); Huntsman (2070Y80R)
Decorative candlesticks: Herbal Walk (5030B90G)
Lacquer serving tray: Lacquer Red
Trompe l'oeil villa : Basket Weave (4050Y40R); Pale Sherbert (0710Y); Mantilla (9000N)

Details of Dulux paint suppliers are available from:

AUSTRALIA
Dulux Australia Ltd
PO Box 60
McNaughton Road
Clayton
Victoria 3168
Australia

CANADA
ICI Paints Canada Ltd
8200 Keele Street
Concord
Ontario
Canada L4K 2A5

GERMANY
ICI Lacke Farben GmbH
Dusseldorfer Str 102
Postfach 940
4010 Hilden
Germany

HONG KONG
ICI Swire Paints Ltd
8/F Luk Kwok Centre
72 Gloucester Raod
Wanchai
Hong Kong

NEW ZEALAND
Dulux New Zealand
PO Box 30-749
Gracefield
Lower Hutt
New Zealand

UK
ICI Paints
Wexham Road
Slough
Berkshire SL2 5DS
England

USA
The Glidden Company
925 Euclid Avenue
Cleveland
Ohio 44115
USA

For further details of Dulux companies worldwide, please write to: ICI Wexham Road, Slough, Berkshire SL2 5DS, England.

ACKNOWLEDGEMENTS

The publishers would like to thank the following companies for providing accessories for use in our photographs.

Bolloms – emulsion and oil glazes; Campbell Marson – wood flooring; Crucial Trading – flooring; Elrose Products – stencils; Forbo Nairn – flooring; General Trading Company – accessories; Harris Brushes – paint brushes; Nice Irma's – accessories; The Pier – accessories; Rowney – oil paints, artist's paints and brushes; Ronseal – wood varnishes; Texas – unpainted wooden furniture; Tintawn – flooring; Joanna Wood – accessories; Dover Street Bookshop – photocopies

GLOSSARY

Some of the paint terms in this book may not be familiar. Here is a list of their U.S. equivalents.

UK	US
cartridge paper	poster paper
cotton wool	cotton balls
dust sheet	drop cloth
emulsion paint	latex
hessian cloth	burlap
lino cuts	linoleum blocks (for printing)
matt finish	flat finish
muslin (cloth)	cheesecloth
mutton-cloth	cheesecloth
paint kettle	paint bucket or can
PVA glue	white craft glue
salt cellar	salt shaker
set square	carpenter's square or framing square
skirting board	baseboard
spirit level	carpenter's level
splatter painting	spatter painting
wardrobe	chiffonrobe or armoire (clothes cupboard)
white spirit	mineral spirits
wire wool	steel wool